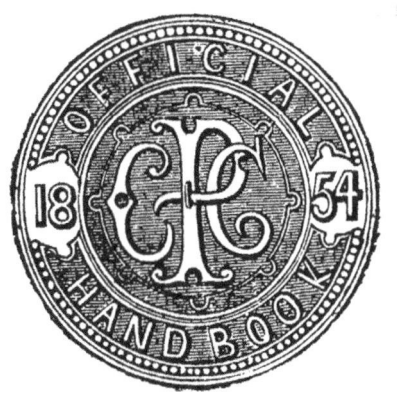

# GEOLOGY
## AND
## INHABITANTS
## OF THE
## ANCIENT WORLD

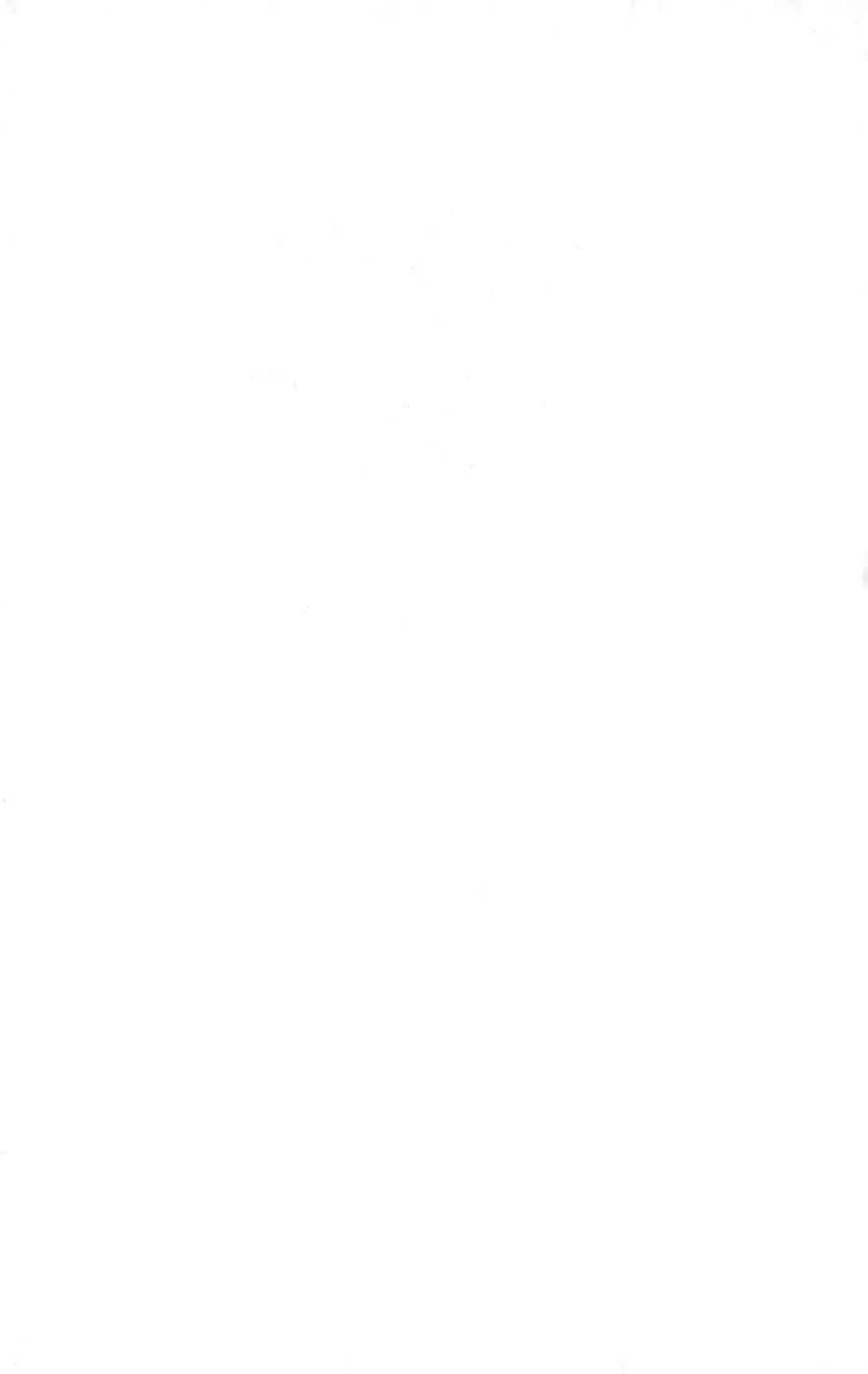

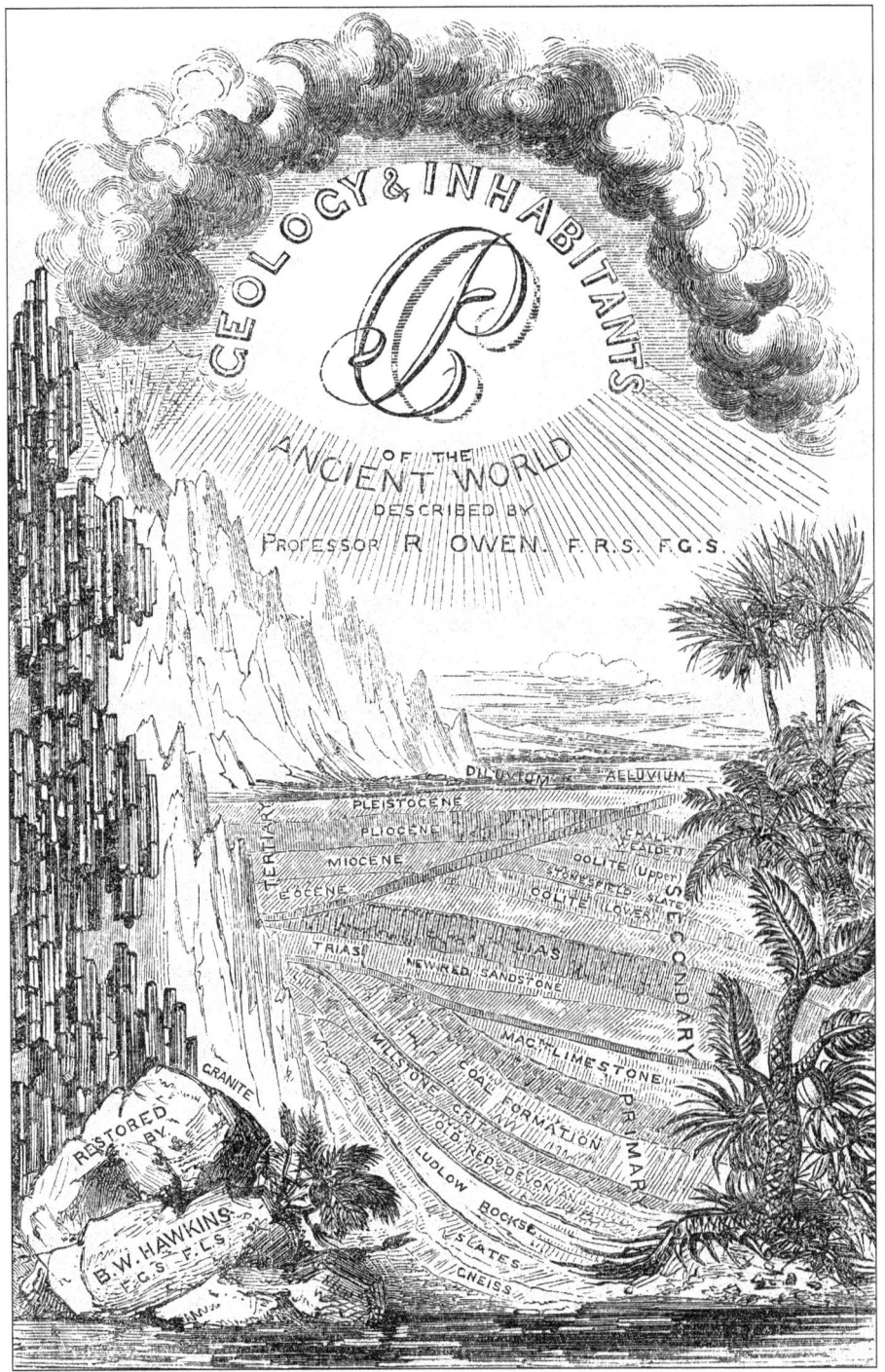

# GEOLOGY AND INHABITANTS

OF THE

# ANCIENT WORLD.

DESCRIBED BY

## RICHARD OWEN, F.R.S.

———◆———

THE ANIMALS CONSTRUCTED BY B. W. HAWKINS, F.G.S.

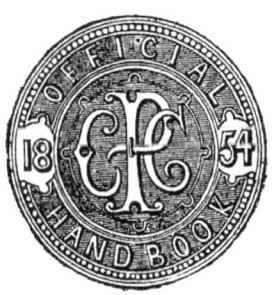

CRYSTAL PALACE LIBRARY,

AND

BRADBURY & EVANS, 11, BOUVERIE STREET, LONDON.

1854.

Euston Grove Press
20 Elderwood Place
London SE27 0HL  UK
www.eustongrove.com

ISBN-13: 978-1-906267-36-0 (paperback)
ISBN-10: 1-906267-36-7 (paperback)
2013 Euston Grove Press

Crystal Palace Guides series number 15

Owen, Richard
*Geology and Inhabitants of the Ancient World*

First published 1854 by Crystal Palace Library
and Bradbury and Evans.

A different edition of this title was published by Euston
Grove Press in 2010 under ISBN 978-1-906267-14-8.

This facsimile edition published in 2013 by Euston Grove
Press.


British Library Cataloguing in Publication Data

A catalogue record for this book is available from the
British Library.

# CONTENTS.

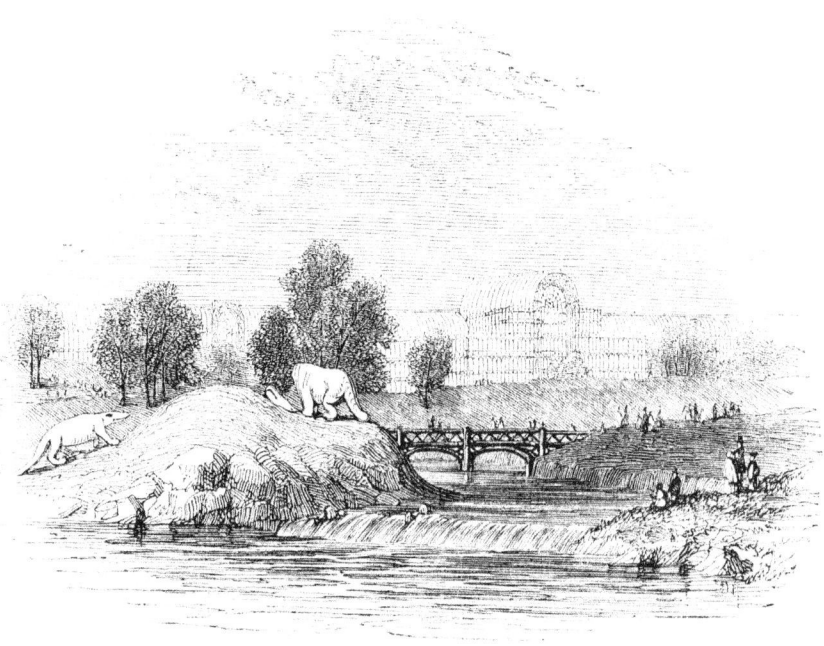

# GEOLOGY AND INHABITANTS OF THE
# ANCIENT WORLD.

### INTRODUCTION.

BEFORE entering upon a description of the restorations of the
Extinct Animals, placed on the Geological Islands in the great
Lake, a brief account may be premised of the principles and pro-
cedures adopted in carrying out this attempt to present a view of
part of the animal creation of former periods in the earth's
history.

Those extinct animals were first selected of which the entire,
or nearly entire, skeleton had been exhumed in a fossil state. To
accurate drawings of these skeletons an outline of the form of the
entire animal was added, according to the proportions and rela-
tions of the skin and adjacent soft parts to the superficial parts of
the skeleton, as yielded by those parts in the nearest allied living

animals. From such an outline of the exterior, Mr. Waterhouse Hawkins prepared at once a miniature model form in clay.

This model was rigorously tested in regard to all its proportions with those exhibited by the bones and joints of the skeleton of the fossil animal, and the required alterations and modifications were successively made, after repeated examinations and comparisons, until the result proved satisfactory.

The next step was to make a copy in clay of the proof model, of the natural size of the extinct animal : the largest known fossil bone, or part, of such animal being taken as the standard according to which the proportions of the rest of the body were calculated agreeably with those of the best preserved and most perfect skeleton. The model of the full size of the extinct animal having been thus prepared, and corrected by renewed comparisons with the original fossil remains, a mould of it was prepared, and a cast taken from this mould, in the material of which the restorations, now exposed to view, are composed.

There are some very rare and remarkable extinct animals of which only the fossil skull and a few detached bones of the skeleton have been discovered : in most of these the restoration has been limited to the head, as, for example, in the case of the Mosasaurus ; and only in two instances—those, viz., of the Labyrinthodon and Dicynodon—has Mr. Hawkins taken upon himself the responsibility of adding the trunk to the known characters of the head, such addition having been made to illustrate the general affinities and nature of the fossil, and the kind of limbs required to produce the impressions of the footprints, where these have been detected and preserved in the petrified sands of the ancient sea-shores trodden by these strange forms of the Reptilian class.

With regard to the hair, the scales, the scutes, and other modifications of the skin, in some instances the analogy of the nearest allied living forms of animals has been the only guide ; in a few instances, as in that of the Ichthyosaurus, portions of the petrified integument have been fortunately preserved, and have guided the artist most satisfactorily in the restoration of the skin and soft parts of the fins ; in the case of other reptiles, the bony plates, spines, and scutes have been discovered in a fossil state, and have been scrupulously copied in the attempt to restore the peculiar tegumentary features of the extinct reptiles, as *e. g.* in the Hylæosaurus.

In every stage of this difficult, and by some it may be thought, perhaps, too bold, attempt to reproduce and present to human gaze and contemplation the forms of animal life that have successively flourished during former geological phases of time, and have passed away long ages prior to the creation of man, the writer of the following brief notice of the nature and affinities of the animals so restored feels it a duty, as it is a high gratification to him, to testify to the intelligence, zeal, and peculiar artistic skill by which his ideas and suggestions have been realised and carried out by the talented director of the fossil department, Mr. Waterhouse Hawkins. Without the combination of science, art, and manual skill, happily combined in that gentleman, the present department of the Instructive Illustrations at the Crystal Palace could not have been realised.

## THE SECONDARY ISLAND.

The most cursory observation of the surface of the earth shows that it is composed of distinct substances, such as clay, chalk, lias, limestone, coal, slate, sandstone, &c. ; and a study of such substances, their relative position and contents, has led to the conviction that these external parts of the earth have acquired their present condition gradually, under a variety of circumstances, and at successive periods, during which many races of animated beings, distinct both from those of other periods and from those now living, have successively peopled the land and the waters ; the remains of these creatures being found buried in many of the layers or masses of mineral substances, forming the crust of the earth.

The object of the Islands in the Geological Lake is to demonstrate the order of succession, or superposition, of these layers or strata, and to exhibit, restored in form and bulk, as when they lived, the most remarkable and characteristic of the extinct animals and plants of each stratum.

The series of mineral substances and strata represented in the smaller island have been called by geologists "secondary formations," because they lie between an older series termed "primary," and a newer series termed " tertiary : " the term " formation " meaning any assemblage of rocks or layers which have some character in common, whether of origin, age, or composition.*

Following the secondary formations as they descend in the earth, or succeed each other from above downwards, and as they are shown, obliquely tilted up out of their original level position from left to right, in the Secondary Island, they consist : 1st, of the Chalk or Cretaceous group ; 2nd, the Wealden ; 3rd, the Oolite ; 4th, the Lias ; and 5th, the New Red Sandstone.

* Lyell, "Manual of Elementary Geology."

## THE CHALK.

THE chalk formations or "cretaceous group of beds" include strata of various mineral substances ; but the white chalk which forms the cliffs of Dover and the adjoining coasts, and the downs and chalk quarries of the South of England, is the chief and most characteristic formation.    Chalk, immense as are the masses in which it has been deposited, owes its origin to living actions ; every particle of it once circulated in the blood or vital juices of certain species of animals, or of a few plants, that lived in the seas of the secondary period of geological time.    White chalk consists of carbonate of lime, and is the result of the decomposition chiefly of coral-animals (*Madrepores, Millepores, Flustra, Cellepora,* &c.), of sea-urchins (*Echini*), and of shell-fishes (*Testacea*), and of the mechanical reduction, pounding, and grinding of their shells. Such chalk-forming beings still exist, and continue their operations in various parts of the ocean, especially in the construction of coral reefs and islands.

Every river that traverses a limestone district carries into the sea a certain proportion of caustic lime in solution : the ill effects of the accumulation of this mineral are neutralised by the power allotted to the above-cited sea-animals to absorb the lime, combine it with carbonic-acid, and precipitate or deposit it in the condition of insoluble chalk, or carbonate of lime.

The entire cretaceous series includes from above downwards :

Maestricht beds of yellowish chalk.

Upper white chalk with flints.

Lower white chalk without flints.

Upper green-sand.

Gault.

Lower green-sand and Kentish rag.

The best known and most characteristic large extinct animal of the chalk formations is chiefly found in the uppermost and most recent division, and is called

## No. 1.—THE MOSASAURUS.

### (*Mosasaurus Hoffmanni*, Hoffmann's Mosasaur.)

Of this animal almost the entire skull has been discovered, but not sufficient of the rest of the skeleton to guide to a complete restoration of the animal. The head only, therefore, is shown, of the natural size, at the left extremity of the Secondary Island.

The first or generic name of this animal is derived from the locality, Maestricht, on the river Meuse (Lat. *Mosa*), in Germany, where its remains have been chiefly discovered, and from the Greek word *sauros*, a lizard, to which tribe of animals it belongs. Its second name refers to its discoverer, Dr. Hoffmann, of Maestricht, surgeon to the forces quartered in that town in 1780. This gentleman had occupied his leisure by the collection of the fossils from the quarries which were then worked to a great extent at Maestricht for a kind of yellowish stone of a chalky nature, and belonging to the most recent of the secondary class of formations in geology. In one of the great subterraneous quarries or galleries, about five hundred paces from the entrance, and ninety feet below the surface, the quarrymen exposed part of the skull of the Mosasaurus, in a block of stone which they were engaged in detaching. On this discovery they suspended their work, and went to inform Dr. Hoffmann, who, on arriving at the spot, directed the operations of the men, so that they worked out the block without injury to the fossil ; and the doctor then, with his own hands, cleared away the matrix and exposed the jaws and teeth, casts of which are shown in the cretaceous rock of the Island.

This fine specimen, which Hoffmann had added with so much pains and care to his collection, soon, however, became a source of chagrin to him. One of the canons of the cathedral at Maestricht, who owned the surface of the soil beneath which was the quarry whence the fossil had been obtained, when the fame of the specimen reached him, pleaded certain feudal rights to it. Hoffmann resisted, and the canon went to law. The Chapter supported the canon, and the decree ultimately went against the poor surgeon, who lost both his specimen and his money—being made to pay the costs of the action.    The canon did not, however, long enjoy possession of the unique specimen. When the French army bombarded Maestricht in 1795, directions were given to spare the suburb in which the famous fossil was known to be preserved ; and after the capitulation of the town it was seized and borne off in triumph.    The specimen has since remained in the museum of the Garden of Plants at Paris.

This skull of the Mosasaurus measures four and a half feet long and two and a half feet wide.    The large pointed teeth on the jaws are very conspicuous ; but, in addition to these, the gigantic reptile had teeth on a bone of the roof of the mouth (the pterygoid), like some of the modern lizards.    The entire length of the animal has been estimated at about thirty feet.    It is conjectured to have been able to swim well, and to have frequented the sea in quest of prey : its dentition shows its predatory and carnivorous character, and its remains have hitherto been met with exclusively in the chalk formations.    Besides the specimens from St. Peter's Mount, Maestricht, of which the above-described skull is the most remarkable, fossil bones and teeth of the Mosasaurus have been found in the chalk of Kent, and in the green-sand—a member of the cretaceous series—in New Jersey, United States of America.    No animal like the Mosasaurus is now known to exist.

### Nos. 2 & 3.—The Pterodactyle.

Nos. 2 and 3 are restorations of a flying reptile or dragon, called Pterodactyle, from the Greek words *pteron*, a wing, and *dactylos*, a finger ; because the wings are mainly supported by the outer finger, enormously lengthened and of proportionate strength, which, nevertheless, answers to the little finger of the human hand. The wings consisted of folds of skin, like the leather wings of the bat ; and the Pterodactyles were covered with scales, not with feathers : the head, though somewhat resembling in shape that of a bird, and supported on a long and slender neck, was provided with long jaws, armed with teeth ; and altogether the structure of these extinct members of the reptilian class is such as to rank them amongst the most extraordinary of all the creatures yet discovered in the ruins of the ancient earth.

Remains of the Pterodactyle were first discovered, in 1784, by Prof. Collini, in the lithographic slate of Aichstadt, in Germany, which slate is a member of the oolitic formations : the species so discovered was at first mistaken for a bird, and afterwards supposed to be a large kind of bat, but had its true reptilian nature demonstrated by Baron Cuvier, by whom it was called the *Pterodactylus longirostris*, or Long-beaked Pterodactyle : it was about the size of a curlew.

A somewhat larger species—the *Pterodactylus macronyx*, or Long-clawed Pterodactyle—was subsequently discovered by the Rev. Dr. Buckland, in the lias formation of Lyme Regis : its wings, when expanded, must have been about four feet from tip to tip. The smallest known species—the *Pterodactylus brevirostris*, or Short-beaked Pterodactyle—was discovered in the lithographic slate at Solenhofen, Germany, and has been described by Professor Soemmering.

Remains of the largest known kinds of Pterodactyle have been discovered more recently in chalk-pits, at Burham, in Kent. The skull of one of these species—the *Pterodactylus Cuvieri*—was about twenty inches in length, and the animal was upborne on an expanse of wing of probably not less than eighteen feet from tip to tip. The restored specimen of this species is numbered 3.

A second very large kind of Pterodactyle—the *Pterodactylus compressirostris*, or Thin-beaked Pterodactyle—had a head from fourteen to sixteen inches in length and an expanse of wing, from tip to tip, of fifteen feet. The remains of this species have also been found in the chalk of Kent. From the same formation and locality a third large kind of Pterodactyle, although inferior in size to the two foregoing, has been discovered, called the *Pterodactylus conirostris*, and also—until the foregoing larger kinds were discovered—*Pterodactylus giganteus*. The long, sharp, conical teeth in the jaws of the Pterodactyles indicate them to have preyed upon other living animals ; their eyes were large, as if to enable them to fly by night. From their wings projected fingers, terminated by long curved claws, and forming a powerful paw, wherewith the animal was enabled to creep and climb, or suspend itself from trees. It is probable, also, that the Pterodactyles had the power of swimming ; some kinds, *e.g.*, the *Pterodactylus Gemmingi*, had a long and stiff tail.    "Thus," writes Dr. Buckland, "like Milton's Fiend, all qualified for all services and all elements, the creature was a fit companion for the kindred reptiles that swarmed in the seas, or crawled on the shores of a turbulent planet.

> ' The Fiend,
> O'er bog, or steep, through strait, rough, dense, or rare,
> With head, hands, wings, or feet, pursues his way,
> And swims, or sinks, or wades, or creeps, or flies.'
> *Paradise Lost*, Book II."

# THE WEALDEN.

THE Wealden is a mass of petrified clay, sand, and sandstone, deposited from the fresh or brackish water of probably some great estuary, and extending over parts of the counties of Kent, Surrey, and Sussex. This fresh-water formation derives its name from the "Weald" or "Wold" of Kent, where it was first geologically studied, and where it is exposed by the removal of the chalk, which covers or overlies it, in other parts of the South of England.

The Wealden is divided into three groups of strata, which succeed each other in the following descending order :—

1st. Weald Clay, sometimes including thin beds of sand and shelly limestone, forming beds of from 140 to 280 feet in depth or vertical thickness.

2nd. Hastings Sand, in which occur some clays and calcareous grits, forming beds of from 400 to 500 feet in depth.

3rd. Purbeck Beds, so called from being exposed chiefly in the Isle of Purbeck, off the coast of Dorsetshire, where it forms the quarries of the limestone for which Purbeck is famous : the beds of limestones and marls are from 150 to 200 feet in depth.

## Nos. 4 & 5.—THE IGUANODON.

### (*Iguanodon Mantelli*, Conybeare.)

One afternoon, in the spring of 1822, an accomplished lady, the wife of a medical practitioner, at Lewes, in Sussex, walking along the picturesque paths of Tilgate Forest, discovered some objects in the coarse conglomerate rock of the quarries of that locality, which, from their peculiar form and substance, she thought would be interesting to her husband, whose attention had been directed, during his professional drives, to the geology and fossils of his neighbourhood.

The lady was Mrs. Mantell : her husband, the subsequently distinguished geologist, Dr. Mantell,* perceived that the fossils discovered by his wife were teeth, and teeth of a large and unknown animal.

" As these teeth," writes the doctor, " were distinct from any that had previously come under my notice, I felt anxious to submit them to the examination of persons whose knowledge and means of observation were more extensive than my own.   I therefore transmitted specimens to some of the most eminent naturalists in this country and on the continent.   But although my communications were acknowledged with that candour and liberality which constantly characterise the intercourse of scientific men, yet no light was thrown upon the subject, except by the illustrious Baron Cuvier, whose opinions will best appear by the following extract from the correspondence with which he honoured me :—

" ' These teeth are certainly unknown to me ; they are not from a carnivorous animal, and yet I believe that they belong, from their slight degree of complexity, the notching of their margins, and the thin coat of enamel that covers them, to the order of reptiles.

" ' May we not here have a new animal !—a herbivorous reptile ? And, just as at the present time with regard to mammals (land-quadrupeds with warm blood), it is amongst the herbivorous that we find the largest species, so also with the reptiles at the remote period when they were the sole terrestrial animals, might not the largest amongst them have been nourished by vegetables ?

" ' Some of the great bones which you possess may belong to this animal, which, up to the present time, is unique in its kind. Time will confirm or confute this idea, since it is impossible but that one day a part of the skeleton, united to portions of jaws with the teeth, will be discovered.' "

" These remarks," Dr. Mantell proceeds to say, " induced me to pursue my investigations with increased assiduity, but hitherto they have not been attended with the desired success, no connected

* " The first specimens of the teeth were found by Mrs. Mantell in the coarse conglomerate of the Forest, in the spring of 1822."—Mantell, "Geology of the South-East of England," 8vo, 1833, p. 268.

portion of the skeleton having been discovered. Among the specimens lately connected, some, however, were so perfect, that I
resolved to avail myself of the obliging offer of Mr. Clift (to whose
kindness and liberality I hold myself particularly indebted), to
assist me in comparing the fossil teeth with those of the recent
Lacertæ in the Museum of the Royal College of Surgeons. The
result of this examination proved highly satisfactory, for in an
Iguana which Mr. Stutchbury had prepared to present to the
College, we discovered teeth possessing the form and structure of
the fossil specimens." (Phil. Trans., 1825, p. 180.) And he
afterwards adds :—" The name Iguanodon, derived from the form
of the teeth, (and which I have adopted at the suggestion of the
Rev. W. Conybeare,) will not, it is presumed, be deemed objec-
tionable." (Ib. p. 184.)

The further discovery which Baron Cuvier's prophetic glance
saw buried in the womb of time, and the birth of which verified
his conjecture that some of the great bones collected by Dr. Mantell
belonged to the same animal as the teeth, was made by Mr. W. H.
Bensted, of Maidstone, the proprietor of a stone-quarry of the
Shanklin-sand formation, in the close vicinity of that town. This
gentleman had his attention one day, in May, 1834, called by his
workmen to what they supposed to be petrified wood in some pieces
of stone which they had been blasting. He perceived that what
they supposed to be wood was fossil bone, and with a zeal and
care which have always characterised his endeavours to secure for
science any evidence of fossil remains in his quarry, he immediately
resorted to the spot. He found that the bore or blast by which these
remains were brought to light, had been inserted into the centre of
the specimen, so that the mass of stone containing it had been shat-
tered into many pieces, some of which were blown into the adjoining
fields. All these pieces he had carefully collected, and proceeding
with equal ardour and success to the removal of the matrix from
the fossils, he succeeded after a month's labour in exposing them
to view, and in fitting the fragments to their proper places.

This specimen is now in the British Museum.

Many other specimens of detached bones, including vertebræ or
parts of the back-bone, especially that part resting on the hind
limbs, and called the " pelvis," bones of the limbs, down to those

that supported the claws, together with jaws and teeth, which have since been successively discovered, have enabled anatomists to re-construct the extinct Iguanodon, and have proved it to have been a herbivorous reptile, of colossal dimensions, analogous to the diminutive Iguana in the form of its teeth, but belonging to a distinct and higher order of reptiles, more akin to the crocodiles. The same rich materials, selecting the largest of the bones as a standard, have served for the present restorations (Nos. 4 and 5) of the animal, as when alive : all the parts being kept in just proportion to the standard bones, and the whole being thus brought to the following dimensions :—

Total length, from the nose or muzzle to the end
of the tail    .    .    .    .    .    34 feet 9 inches.
Greatest girth of the trunk    .    .    .    20  ,,  5  ,,
Length of the head    .    .    .    .    3  ,,  6  ,,
Length of the tail    .    .    .    .    .    15  ,,  6  ,,

The character of the scales is conjectural, and the horn more than doubtful, though attributed to the Iguanodon by Dr. Mantell and most geologists.

This animal probably lived near estuaries and rivers, and may have derived its food from the *Clathrariæ, Zamiæ, Cycades,* and other extinct trees, of which the fossil remains abound in the same formations as those yielding the bones and teeth of the Iguanodon.

These formations are the Wealden and the Neocomian or green-sand : the localities in which the remains of the Iguanodon have been principally found, are the Weald of Kent and Sussex : Horsham, in Sussex ; Maidstone, in Kent ; and the Isle of Wight.

Restorations of the *Cycas* and *Zamia* are placed, with the Iguanodon, on the Wealden division of the Secondary Island.


No. 6.—The Hylæosaurus. (*Hylæosaurus Owenii.*)

The animal, so called by its discoverer, Dr. Mantell, belongs to the same highly organised order of the class of reptiles as the Iguanodon, that, viz., which was characterised by a longer and stronger sacrum and pelvis, and by larger limbs than the reptiles of the present day possess ; they were accordingly better fitted for progression on dry land, and probably carried their body higher and more freely above the surface of the ground.

Visiting, in the summer of 1832, a quarry in Tilgate Forest, Dr. Mantell had his attention attracted to some fragments of a large mass of stone, which had recently been broken up, and which exhibited traces of numerous pieces of bone. The portions of the rock, which admitted of being restored together, were cemented, and then the rock was chiselled from the fossil bones, which consisted of part of the back-bone or vertebral column, some ribs, the shoulder bones called scapula and coracoid, and numerous long angular bones or spines which seemed to have supported a lofty serrated or jagged crest, extended along the middle of the back, as in some of the small existing lizards, *e.g.*, the Iguana : cut No. 6. Many small dermal bones were also found, which indicate the Hylæosaurus to have been covered by hard tuberculate scales, like those of some of the Australian lizards, called *Cyclodus*.

This character of the skin, and the serrated crest, are accurately given in the restoration, the major part of which, however, is necessarily at present conjectural, and carried out according to the general analogies of the saurian form. The size is indicated with more certainty according to the proportions of the known vertebræ and other bones.

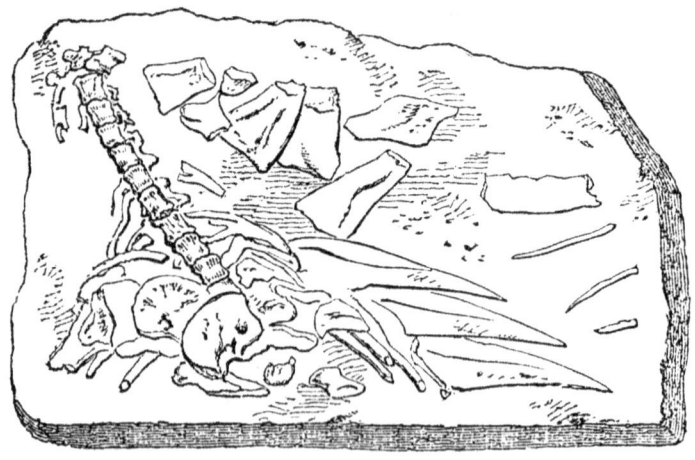

No. 6. Diagram of the Slab containing the Bones of Hylæosaurus.

# THE OOLITE.

THE division of the secondary formations, called " Oolite," takes its name from the most characteristic of its constituents, which is a variety of limestone composed of numerous small grains, resembling the "roe" or eggs of a fish, whence the term, (from the Greek *oon*, an egg, *lithos*, a stone). The oolite, however, includes a great series of beds of marine origin, which, with an average breadth of thirty miles, extend across England, from Yorkshire in the north-east to Dorsetshire in the south-west.

The oolite series lies below the Wealden, and where this is wanting, below the chalk, and consists of the following subdivisions, succeeding each other in the descending order :—

OOLITE.

Upper. { Portland stone and sand.
{ Kimmeridge clay.

Middle. { Coral rag.
{ Oxford clay.

Lower. { Cornbrash and forest marble.
{ Great oolite and Stonesfield slate.
{ Fuller's earth.
{ Inferior oolite.

Upon the portion of the island representing the oolite series, the most conspicuous of the restored animals of that period is—

No. 7.—THE MEGALOSAURUS.

The Megalosaurus, as its name implies (compounded by its discoverer, Dr. Buckland, from the Greek *megas*, great, and *sauros*, lizard), was a lizard-like reptile of great size, " of which," writes Dr. Buckland, "although no skeleton has yet been found entire, so many perfect bones and teeth have been discovered in the same quarries, that we are nearly as well acquainted with the form and dimensions of the limbs as if they had been found together in a single block of stone."

The restoration of the animal has been accordingly effected, agreeably with the proportions of the known parts of the skeleton, and in harmony with the general characters of the order of reptiles to which the **Megalosaurus** belonged. This order—the Dinosauria (Gr. *deinos*, terribly great *sauros*, a lizard)—is that to which the two foregoing huge reptiles of the Wealden series belong, viz., the Iguanodon and Hylæosaurus, and is characterised

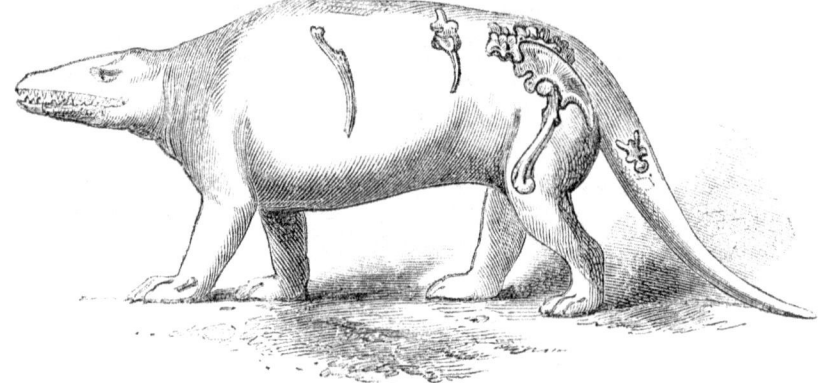

No. 7. Megalosaurus.

by the modifications already mentioned, that fitted them for more efficient progression upon dry land. The **Iguanodon** represented the herbivorous section of the order, the **Hylæosaurus** appears, from its teeth, to have been a mixed feeder, but the Megalosaurus was decidedly carnivorous, and, probably, waged a deadly war against its less destructively endowed congeners and contemporaries.

Baron Cuvier estimated the Megalosaurus to have been about fifty feet in length ; my own calculations, founded on more complete evidence than had been at the Baron's command, reduce its size to about thirty-five feet :* but with the superior proportional height and capacity of trunk, as contrasted with the largest existing crocodiles, even that length gives a most formidable character to this extinct predatory reptile.

As the thigh-bone (*femur*) and leg-bone (*tibia*) measure each nearly three feet, the entire hind-leg, allowing for the cartilages of the joints, must have attained a length of two yards : a bone of the

* "Report of British Fossil Reptiles," 1841, p. 110.

foot (metatarsal) thirteen inches long, indicates that part, with the toes and claws entire, to have been at least three feet in length. The form of the teeth shows the Megalosaurus to have been strictly carnivorous, and viewed as instruments for providing food for so enormous a reptile, the teeth were fearfully fitted to the destructive office for which they were designed. They have compressed conical sharp-pointed crowns, with cutting and finely serrated anterior and posterior edges ; they appear straight, as seen when they had just protruded from the socket, but become bent slightly backwards in the progress of growth, and the fore part of the crown, below the summit, becomes thick and convex.

A minute and interesting description of these teeth will be found in Dr. Buckland's admirable " Bridgewater Treatise " (vol. i. p. 238), from which he concludes that the teeth of the Megalosaurus present " a combination of contrivances analogous to those which human ingenuity has adopted in the construction of the knife, the sabre, and the saw." The fossils which brought to light the former existence of this most formidable reptile, were discovered in 1823, in the oolitic slate of Stonesfield, near Oxford, and were described by Dr. Buckland, in the volume of the " Geological Transactions" for the year 1824.

Remains of the Megalosaurus have since been discovered in the " Bath oolite," which is immediately below the Stonesfield slate, and in the " Cornbrash," which lies above it. Vertebræ, teeth, and some bones of the extremities have been discovered in the Wealden of Tilgate Forest, Kent, and in the ferruginous sand, of the same age, near Cuckfield, in Sussex. Remains of the Megalosaurus also occur in the Purbeck limestone at Swanage Bay, and in the oolite in the neighbourhood of Malton, in Yorkshire.

Mr. Waterhouse Hawkins's restoration, according to the proportions calculated from the largest portions of fossil bones of the Megalosaurus hitherto obtained, yields a total length of the animal, from the muzzle to the end of the tail, of thirty-seven feet ; the length of the head being five feet, the length of the tail fifteen feet ; and the greatest girth of the body twenty-two feet six inches.

## Nos. 8 & 9.—Pterodactyles of the Oolite.

To the right of the Hylæosaurus, on the rock representing the greater oolite formation, are restorations of species of Pterodactyle (*Pterodactylus Bucklandi*, No. 9), smaller than and distinct from those of the chalk formations. The remains of Buckland's Pterodactyle are found pretty abundantly in the oolitic slate of Stonesfield, near Oxford.

## Nos. 10 & 11.—Teleosaurus.

On the shore beneath the overhanging cliff of oolitic rock are two restorations, Nos. 10 and 11, of a large extinct kind of crocodile, to which the long and slender-jawed crocodile of the Ganges, called "Gaviàl" or "Gharriàl" by the Hindoos, offers the nearest resemblance at the present day. Remains of the ancient extinct British gavials have been found in most of the localities where the oolitic formations occur, and very abundantly in the lias cliffs near Whitby, in Yorkshire. The name Teleosaurus (*telos*, the end, *sauros*, a lizard), was compounded from the Greek by Professor Geoffroy St. Hilaire, for a species of these fossil gavials, found by him in the oolite stone at Caen, in Normandy, and has reference to his belief that they formed one—the earliest—extreme of the crocodilian series, as this series has been successively developed in the course of time on our planet.

The jaws are armed with numerous long, slender, sharp-pointed, slightly curved teeth, indicating that they preyed on fishes, and the young or weaker individuals of co-existing reptiles. The nostril is situated more at the end of the upper jaw than in the modern gavial : the fore-limbs are shorter, and the hind ones longer and stronger than in the gavial, which indicates that the Teleosaur was a better swimmer ; the vertebræ or bones of the back are united by slightly concave surfaces, not interlocked by cup and ball joints as in the modern crocodiles, whence it would seem that the Teleosaur lived more habitually in the water, and less seldom moved on dry land ; and, as its fossil remains have been hitherto found only in the sedimentary deposits from the sea, it may be inferred that it was more strictly marine than the crocodile of the Ganges.

The first specimen of a Teleosaur that was brought to light was from the "alum-schale" which forms one layer of the lofty lias cliffs of the Yorkshire coast, near Whitby. A brief description, and figures, of this incomplete fossil skeleton were published by Messrs. Wooller and Chapman, in separate communications, in the 50th volume of the "Philosophical Transactions," in 1758. Captain Chapman observes, "it seems to have been an alligator;" and Mr. Wooller thought "it resembled in every respect the Gangetic gavial." Thus, nearly a century ago, the true nature of the fossil was almost rightly understood, and various were the theories then broached to account for the occurrence of a supposed Gangetic reptile in a petrified state in the cliffs of Yorkshire. It has required the subsequent progress of comparative anatomy to determine, as by the characters above defined, the essential distinction of the Teleosaur from all known existing forms of crocodilian reptiles.

Very abundant remains, and several species, of the extinct genus have been subsequently discovered : but always in the oolitic and liassic formations of the secondary series of rocks.

The oolitic group of rocks are very rich in remains of both plants and animals : many reptiles of genera and species distinct from those here restored have been recognised and determined by portions of the skeleton. Extremely numerous are the remains of fishes, chiefly of an almost extinct order (*Ganoidei*), characterised by hard, shining, enamelled scales. But the most remarkable fossils are those which indisputably prove the existence, during the period of the "Great" or "Lower Oolite," of insectivorous and marsupial mammalia—*i.e.*, of warm-blood quadrupeds, which, like the shrew or hedgehog, fed on insects, and, like the opossum, had a pouch for the transport of the young. The lower jaw of one of these earliest known examples of the mammalian class, found in the Stonesfield slate, near Oxford, may be seen at the British Museum, to which it was presented by J. W. Broderip, Esq., F.R.S., by whom it was described in the "Zoological Journal," vol. iii., p. 408.

It is interesting to observe that the marsupial genera, to which the above fossil quadruped, called *Phascolotherium*, was most nearly allied, are now confined to New South Wales and Van

Diemen's Land ; since it is in the Australian seas that is found the *Cestracion,* a cartilaginous fish which has teeth that are most like those fossil teeth called *Acrodus* and *Psammodus,* so common in the oolite. In the same Australian seas, also, near the shore, the beautiful shell-fish called *Trigonia* is found living, of which genus many fossil species occur in the Stonesfield slate. Moreover, the Araucarian pines are now abundant, together with ferns, in Australia, as they were in Europe in the oolitic period.

# THE LIAS.

"Lias" is an English provincial name adopted in geology, and applied to a formation of limestone, marl, and petrified clay, which forms the base of the oolite, or immediately underlies that division of secondary rocks. The lias has been traced throughout a great part of Europe, forming beds of a thickness varying from 500 to 1000 feet of the above-mentioned substances, which have been gradually deposited from a sea of corresponding extent and direction. The lias abounds with marine shells of extinct species, and with remains of fishes that were clad with large and hard shining scales. Of the higher or air-breathing animals of that period, the most characteristic were the

## ENALIOSAURIA.

The creatures called Enaliosauria or Sea-lizards (from the Greek *enalios*, of the sea, and *sauros*, lizard), were vertebrate animals, or had back bones, breathed the air like land quadrupeds, but were cold-blooded, or of a low temperature, like crocodiles and other reptiles. The proof that the Enaliosaurs respired atmospheric air immediately, and did not breathe water by means of gills like fishes, is afforded by the absence of the bony framework of the gill apparatus, and by the presence, position, and structure of the air passages leading from the nostrils, and also by the bony mechanism of the capacious chest or thoracic-abdominal cavity : all of which characters have been demonstrated by their fossil skeletons. With these characters the Sea-lizards combined the presence of two pairs of limbs shaped like fins, and adapted for swimming.

The Enaliosauria offer two principal modifications of their anatomical, and especially their bony, structure, of which the two kinds grouped together under the respective names of Ichthyosaurus and Plesiosaurus are the examples.

## THE ICHTHYOSAURUS.

The genus Ichthyosaurus includes many species : of which three of the best known and most remarkable have been selected for restoration to illustrate this most singular of the extinct forms of animal life.

The name (from the Greek *ichthys*, a fish, and *sauros*, a lizard) indicates the closer affinity of the Ichthyosaur, as compared with the Plesiosaur, to the class of fishes. The Ichthyosaurs are remarkable for the shortness of the neck and the equality of the width of the back of the head with the front of the chest, impressing the observer of the fossil skeleton with a conviction that the ancient animal must have resembled the whale tribe and the fishes in the absence of any intervening constriction or "neck."

This close approximation in the Ichthyosaurs to the form of the most strictly aquatic back-boned (vertebrate) animals of the existing creation is accompanied by an important modification of the surfaces forming the joints of the back-bone, each of which surfaces is hollow, leading to the inference that they were originally connected together by an elastic bag, or "capsule," filled with fluid— a structure which prevails in the class of fishes, but not in any of the whale or porpoise tribe, nor in any, save a few of the very lowest and most fish-like, of the existing reptiles.

With the above modifications of the head, trunk, and limbs, in relation to swimming, there co-exist corresponding modifications of the tail. The bones of this part are much more numerous than in the Plesiosaurs, and the entire tail is consequently longer ; but it does not show any of those modifications that characterise the bony support of the tail in fishes. The numerous "caudal vertebræ" of the Ichthyosaurus gradually decrease in size to the end of the tail, where they assume a compressed form, or are flattened from side to side, and thus the tail instead of being short and broad, as in fishes, is lengthened out as in crocodiles.

The very frequent occurrence of a fracture of the tail, about one fourth of the way from its extremity, in well-preserved and entire fossil skeletons, is owing to that proportion of the end of the tail having supported a tail-fin. The only evidence which the fossil skeleton of a whale would yield of the powerful horizontal tail-fin characteristic of the living animal, is the depressed or horizontally flattened form of the bones supporting such fin. It is inferred, therefore, from the corresponding bones of the Ichthyosaurus being flattened from side to side, that it possessed a tegumentary tail-fin expanded in the vertical direction. The shape of a fin composed of such perishable material is of course conjectural, but from analogies, not necessary here to further enlarge upon, it was probably like, or nearly like, that which the able artist engaged in the restoration of the entire form of the animal has given to it. Thus, in the construction of the principal swimming-organ of the Ichthyosaurus we may trace, as in other parts of its structure, a combination of mammalian (beast-like), saurian (lizard-like), and piscine (fish-like) peculiarities. In its great length and gradual diminution we perceive its saurian character ; the tegumentary nature of the fin, unsustained by bony fin-rays, bespeaks its affinity to the same part in the mammalian whales and porpoises ; whilst its vertical position makes it closely resemble the tail-fin of the fish.

The horizontality of the tail-fin of the whale tribe is essentially connected with their necessities as warm-blooded animals breathing atmospheric air ; without this means of displacing a mass of water in the vertical direction, the head of the whale could not be brought with the required rapidity to the surface to respire ; but the Ichthyosaurs, not being warm-blooded, or quick breathers, would not need to bring their head to the surface so frequently, or so rapidly, as the whale ; and, moreover, a compensation for the want of horizontality of their tail-fin was provided by the addition of a pair of hind-paddles, which are not present in the whale tribe.    The vertical fin was a more efficient organ in the rapid cleaving of the liquid element, when the Ichthyosaurs were in pursuit of their prey, or escaping from an enemy.

That the Ichthyosaurs occasionally sought the shores, crawled on the strand, and basked in the sunshine, may be inferred from the bony structure connected with their fore-fins, which does not exist in any porpoise, dolphin, grampus, or whale ; and for want of which, chiefly, those warm-blooded, air-breathing, marine animals are so helpless when left high and dry on the sands : the structure in question in the Ichthyosaur is a strong osseous arch, inverted and spanning across beneath the chest from one shoulder-joint to the other ; and what is most remarkable in the structure of this "scapular" arch, as it is called, is, that it closely resembles, in the number, shape, and disposition of its bones, the same part in the singular aquatic mammalian quadruped of Australia, called *Ornithorhynchus, Platypus,* and Duck-mole. The Ichthyosaurs, when so visiting the shore, either for sleep, or procreation, would lie, or crawl prostrate, or with the belly resting or dragging on the ground.

The most extraordinary feature of the head was the enormous magnitude of the eye ; and from the quantity of light admitted by the expanded pupil it must have possessed great powers of vision, especially in the dusk. It is not uncommon to find in front of the orbit (cavity for the eye), in fossil skulls, a circular series of petrified thin bony plates, ranged round a central aperture, where the pupil of the eye was placed. The eyes of many fishes are defended by a bony covering consisting of two pieces ; but a compound circle of overlapping plates is now found only in the eyes of turtles, tortoises, lizards, and birds. This curious apparatus of bony plates would aid in protecting the eyeball from the waves of the sea when the Ichthyosaurus rose to the surface, and from the pressure of the dense element when it dived to great depths ; and they show, writes Dr. Buckland,* " that the enormous eye, of which they formed the front, was an optical instrument of varied and prodigious power, enabling the Ichthyosaurus to descry its prey at great or little distances, in the obscurity of night, and in the depths of the sea."

Of no extinct reptile are the materials for a complete and exact restoration more abundant and satisfactory than of the Ichthyosaurus they plainly show that its general external figure must have been that of a huge predatory abdominal fish, with a longer tail, and a smaller tail-fin : scale-less, moreover, and covered by a smooth, or finely wrinkled skin analogous to that of the whale tribe.

The mouth was wide, and the jaws long, and armed with numerous pointed teeth, indicative of a predatory and carnivorous nature in all the species ; but these differed from one another in regard to the relative strength of the jaws, and the relative size and length of the teeth.

Masses of masticated bones and scales of extinct fishes, that lived in the same seas and at the same period as the Ichthyo-

* Op. cit., p. 174.

saurus, have been found under the ribs of fossil specimens, in the situation where the stomach of the animal was placed ; smaller, harder, and more digested masses, containing also fish-bones and scales have been found, bearing the impression of the structure of the internal surface of the intestine of the great predatory sea-lizard.    These digested masses are called " coprolites."

In tracing the evidences of creative power from the earlier to the later formations of the earth's crust, remains of the Ichthyosaurus are first found in the lower lias, and occur, more or less abundantly, through all the superincumbent secondary strata up to, and inclusive of, the chalk formations.    They are most numerous in the lias and oolite, and the largest and most characteristic species have been found in these formations.

### No. 12.—ICHTHYOSAURUS PLATYODON.

This most gigantic species, so called on account of the crown of the tooth being more flattened than in other species, and having sharp edges, as well as a sharp point, was first discovered in the lias of Lyme Regis, in Dorsetshire.    Fossil remains now in the British Museum, and in the museum of the Geological Society, fully bear out the dimensions exhibited by the restoration of the animal as seen basking on the shore between the two specimens of Long-necked Plesiosaurs.    The head of this species is relatively larger in proportion to the trunk, than in the *Ichthyosaurus communis* or *Ichthyosaurus tenuirostris:* the lower jaw is remarkably massive and powerful, and projects backwards beyond the joint, as far as it does in the crocodile.    In the skull of an individual of this species, preserved in the apartments of the Geological Society of London, the cavity for the eye, or orbit, measures, in its long diameter, fourteen inches.    The fore and hind paddles are large and of equal size.

The lias of the valley of Lyme Regis, Dorsetshire, is the chief grave-yard of the *Ichthyosaurus platyodon ;* but its remains are pretty widely distributed.    They have been found in the lias of Glastonbury, of Bristol, of Scarborough and Whitby, and of Bitton, in Gloucestershire ; some vertebræ, apparently of this species, have likewise been found in the lias at Ohmden, in Germany.

## No. 13.—Ichthyosaurus tenuirostris.

Behind the *Ichthyosaurus platyodon,* is placed the restoration of the *Ichthyosaurus tenuirostris,* or Slender-snouted Fish-lizard. The most striking peculiarity of this species is the great length and slenderness of the jaw-bones, which, in combination with the large eye-sockets and flattened cranium, give to the entire skull a form which resembles that of a gigantic snipe or woodcock, with the bill armed with teeth. These weapons, in the present species, are relatively more numerous, smaller, and more sharply pointed than in the foregoing, and indicate that the *Ichthyosaurus tenuirostris* preyed on a smaller kind of fish. The fore-paddles are larger than the hind ones. In the museum of the Philosophical Institution, at Bristol, there is an almost entire skeleton of the present species which measures thirteen feet in length. It was discovered in the lias of Lyme Regis. Portions of jaws and other parts of the skeletons of larger individuals have been found fossil in the liasnear Bristol, at Barrow-on-Soar, in Leicestershire, and at Stratford-on-Avon. The *Ichthyosaurus tenuirostris* has also left its remains in the lias formation at Boll and Amburg, in Wirtemberg, Germany.

## No. 14.—Ichthyosaurus communis.

Of this species, which was the most "common," when first discovered in 1824, but which has since been surpassed by other species in regard to the known number of individuals, the head is restored, as protruded from the water, to the right of the foregoing species.

The *Ichthyosaurus communis* is characterised by its relatively large teeth, with expanded, deeply-grooved bases, and round conical furrowed crowns ; the upper jaw contains, on each side, from forty to fifty of such teeth. The fore-paddles are three times larger than the hind ones. With respect to the size which it attained, the *Ichthyosaurus communis* seems only to be second to the *Ichthyosaurus platyodon*. In the museum of the Earl of Enniskillen, there is a fossil skull of the *Ichthyosaurus communis* which measures, in length, two feet nine inches, indicating an animal of at least twenty feet in length.

## PLESIOSAURUS.

The discovery of this genus forms one of the most important additions that geology has made to comparative anatomy.    Baron Cuvier deemed " its structure to have been the most singular, and its characters the most monstrous, that had been yet discovered amid the ruins of a former world." To the head of a lizard it united the teeth of a crocodile, a neck of enormous length, resembling the body of a serpent, a trunk and tail having the proportions of an ordinary quadruped, the ribs of a chameleon, and the paddles of a whale. "Such," writes Dr. Buckland, "are the strange combinations of form and structure in the Plesiosaurus, a genus, the remains of which, after interment for thousands of years amidst the wreck of millions of extinct inhabitants of the ancient earth, are at length recalled to light by the researches of the geologist, and submitted to our examination, in nearly as perfect a state as the bones of species that are now existing upon the earth." (Op. cit., vol. v. p. 203).

The first remains of this animal were discovered in the lias of Lyme Regis, about the year 1823, and formed the subject of the paper by the Rev. Mr. Conybeare (now Dean of Llandaff), and Mr. (now Sir Henry) De la Beche, in which the genus was established and named Plesiosaurus (from the Greek words, *plesios* and *sauros*, signifying "near" or "allied to," and "lizard"), because the authors saw that it was more nearly allied to the lizard than was the Ichthyosaurus from the same formation.

The entire and undisturbed skeletons of several individuals, of different species, have since been discovered, fully confirming the sagacious restorations by the original discoverers of the *Plesiosaurus*. Of these species three have been selected as the subjects of Mr. Waterhouse Hawkins's reconstructions and representations of the living form of the strange reptiles.

### No. 15.—PLESIOSAURUS MACROCEPHALUS.

The first of these has been called, from the relatively larger size of the head, the *Plesiosaurus macrocephalus* (No. 15), (Gr. *macros*, long, *cephale*, head). The entire length of the animal, as indicated by the largest remains, and as given in the restoration, is eighteen

feet, the length of the head being two feet, that of the neck six
feet ; the greatest girth of the body yields seven feet.

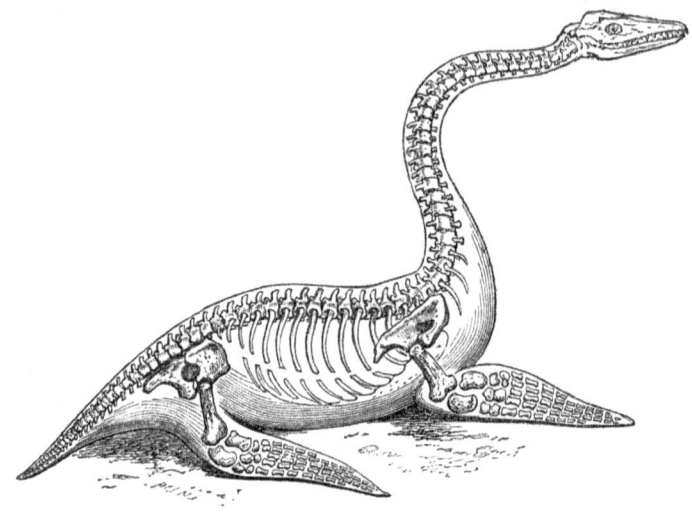

No. 15. Plesiosaurus macrocephalus.

Although Baron Cuvier and Dr. Buckland both rightly allude
to the resemblance of the fins or paddles of the Plesiosaur to
those of the whale, yet this most remarkable difference must be
borne in mind, that, whereas the whale tribe have never more than
one pair of fins, the Plesiosaurs have always two pairs, answering
to the fore and hind limbs of land quadrupeds ; and the fore-pair
of fins, corresponding to those in the whale, differed by being more
firmly articulated, through the medium of collar-bones (clavicles),
and of two other very broad and strong bones (called coracoids),
to the trunk (thorax), whereby they were the better enabled to
move the animal upon dry land.

Remains of the *Plesiosaurus macrocephalus* have been discovered
in the lias of Lyme Regis, in Dorsetshire, and of Weston, in
Somersetshire.

### No. 16.—PLESIOSAURUS DOLICHODEIRUS.

Further to the left, on the shore of the Secondary Island, is a
restoration of the *Plesiosaurus dolichodeirus*, or Long-necked Plesiosaurus (No. 16). The head in this remarkable species is smaller, and

the neck proportionally longer than in the *Plesiosaurus macrocephalus*. The remains of the Long-necked Plesiosaur have been found chiefly at Lyme Regis, in Dorsetshire. The well known specimen of an almost entire skeleton, formerly in the possession of His Grace the Duke of Buckingham, is now in the British Museum.

## No. 17.—PLESIOSAURUS HAWKINSII.

The most perfect skeletons of the Plesiosaurus are those that have been wrought out of the lias at Street, near Glastonbury, by Mr. Thomas Hawkins, F.G.S., and which have been purchased by the trustees of the British Museum. A restoration is given by Mr. Waterhouse Hawkins, at No. 17, of a species with characters somewhat intermediate between the Large-headed and Long-necked Plesiosaurs, and which has been called, after its discoverer, *Plesiosaurus Hawkinsii*.

The Plesiosaurs breathed air like the existing crocodiles and the whale tribe, and appear to have lived in shallow seas and estuaries. That the Long-necked Sea-lizard was aquatic is evident from the form of its paddles; and that it was marine is almost equally so, from the remains with which its fossils are universally associated; that it may have occasionally visited the shore, the resemblance of its extremities to those of a turtle leads us to conjecture; its motion, however, must have been very awkward on land; its long neck must have impeded its progress through the water, presenting a striking contrast to the organisation which so admirably adapted the Ichthyosaurus to cut its swift course through the waves. "May it not, therefore, be concluded that it swam upon, or near the surface," asks its accomplished discoverer, "arching back its long neck like a swan, and occasionally darting it down at the fish that happened to float within its reach? It may perhaps have lurked in shoal-water along the coast, concealed among the sea-weed, and, raising its nostrils to a level with the surface from a considerable depth, may have found a secure retreat from the assaults of dangerous enemies; while the length and flexibility of its neck may have compensated for the want of strength in its jaws, and its incapacity for swift motion through the water, by the sudden-

D

ness and agility of the attack which it enabled it to make on every animal fitted for its prey which came within its reach." *

For the Secondary Island three species of the Plesiosaurus have been restored, the *Plesiosaurus macrocephalus*, the *Plesiosaurus dolichodeirus* (Gr. *dolichos*, long, *deire*, neck), and the *Plesiosaurus Hawkinsii*. The name "long-necked" was given to the second of these species before it was known that many other species with long and slender necks had existed in the seas of the same ancient period : the third species is named after Mr. Thomas Hawkins, F.G.S., the gentleman by whose patience, zeal, and skill, the British Museum has been enriched with so many entire skeletons of these most extraordinary extinct sea-lizards.

The remains of all these species occur in the lias at Lyme Regis, and at Street, near Glastonbury ; but the *Plesiosaurus Hawkinsii* is the most abundant in the latter locality.

* "Transactions of the Geological Society," Second Series, vi. 503.    1841.

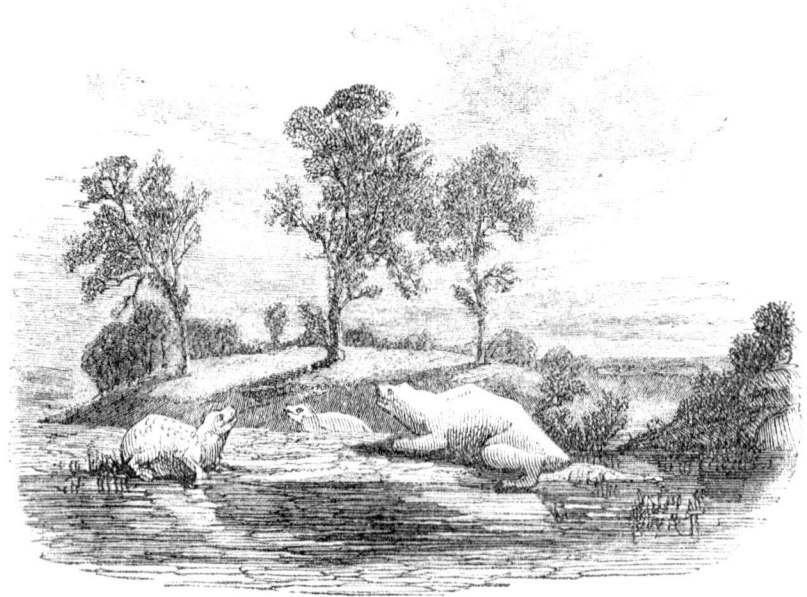

# NEW RED SANDSTONE.

"Trias" is an arbitrary term applied in geology to the upper division of a vast series of red loams, shales, and sandstones, interposed between the lias and the coal, in the midland and western counties of England. This series is collectively called the "New Red Sandstone formation," to distinguish it from the "Old Red Sandstone formation," of similar or identical mineral character, which lies immediately beneath the coal.

The animals which have been restored and placed on the lowest formation of the Secondary Island, are peculiar to the "triassic," or upper division of the "New Red Sandstone" series, which division consists, in England, of saliferous (salt-including) shales and sandstones, from 1000 to 1500 feet thick in Lancashire and Cheshire, answering to the formation called "Keuper-sandstone" by the German geologists ; and of sandstone and quartzose conglomerate of 600 feet in thickness, answering to the German "Bunter-sandstone."

The largest and most characteristic animals of the trias are reptiles of the order

### BATRACHIA.

The name of this order is from the Greek word *batrachos*, signifying a frog : and the order is represented in the present animal-population of England by a few diminutive species of frogs, toads, and newts, or water-salamanders. But, at the period of the deposition of the new red sandstone, in the present counties of Warwick and Cheshire, the shores of the ancient sea, which were then formed by that sandy deposit, were trodden by reptiles, having the essential bony characters of the Batrachia, but combining these with other bony characters of crocodiles and lizards ; and exhibiting both under a bulk which is made manifest by the restoration of the largest known species, (No. 16), occupying the extreme promontory of the Island, illustrative of the lowest and oldest deposits of the secondary series of rocks. The species in question is called the—

### No. 18.—LABYRINTHODON SALAMANDROIDES.

or the Salamander-like Labyrinthodon ; the latter term being from the Greek, signifying the peculiar structure of the teeth, which differ from all other reptiles in the huge Batrachia in question, by reason of the complex labyrinthic interblending of the different substances composing the teeth. The skull of the Labyrinthodon is attached to the neck-bones by two joints or condyles, and the teeth are situated both on the proper jaw-bones, and on the bone of the roof of the mouth called " vomer :" both these characters are only found at the present day in the frogs and salamanders.

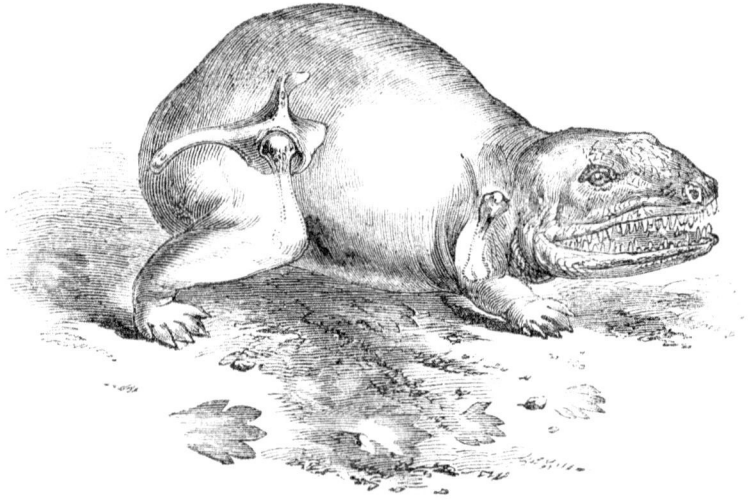

No. 18. Labyrinthodon Salamandroides.

The hind-foot of the Labyrinthodon was also, as in the toad and frog, much larger than the fore-foot ; and the innermost digit in both was short and turned in, like a thumb.

Consecutive impressions of the prints of these feet have been traced for many steps in succession (as is accurately represented in the new red sandstone part of the Secondary Island) in quarries of that formation in Warwickshire, Cheshire, and also in Lancashire, more especially at a quarry of a whitish quartzose sandstone at Storton Hill, a few miles from Liverpool. The foot-marks are partly concave and partly in relief ; the former are seen upon the upper surface of the sandstone slabs, but those in relief are only upon the lower surfaces, being, in fact, natural casts, formed on the subjacent foot-prints as in moulds. The impressions of the hind-foot are generally eight inches in length and five inches in width : near each large footstep, and at a regular distance—about an inch and a half—before it, a smaller print of the fore-foot, four inches long and three inches wide, occurs. The footsteps follow each other in pairs, each pair in the same line, at intervals of about fourteen inches from pair to pair. The large as well as the small steps show the thumb-like toe alternately on the right and left side, each step making a print of five toes.

Foot-prints of corresponding form but of smaller size have been discovered in the quarry at Storton Hill, imprinted on five thin beds of clay, lying one upon another in the same quarry, and separated by beds of sandstone. From the lower surface of the sandstone layers, the solid casts of each impression project in high relief, and afford models of the feet, toes, and claws of the animals which trod on the clay.

Similar foot-prints were first observed in Saxony, at the village of Hessberg, near Hillburghausen, in several quarries of a gray quartzose sandstone, alternating with beds of red sandstone, and of the same geological age as the sandstones of England that had been trodden by the same strange animal. The German geologist, who first described them, proposed the name of *Cheirotherium* (Gr. *cheir*, the hand, *therion*, beast), for the great unknown animal that had left the foot-prints, in consequence of the resemblance, both of the fore and hind feet, to the impression of a human hand, and Dr. Kaup conjectured that the animal might be a large species of the opossum-kind. The discovery, however, of fossil skulls, jaws, teeth, and a few other bones in the sandstones exhibiting the footprints in question, has rendered it more probable that both the footprints and the fossils are evidences of the same kind of huge extinct Batrachian reptiles.

An entire skull of the largest species discovered in the new red sandstones of Wurtemberg ; a lower jaw of the same species found in the same formation in Warwickshire ; some vertebræ, and a few fragments of bones of the limbs, have served, with the indications of size and shape of the trunk of the animal yielded by the series of consecutive foot-prints, as the basis of the restoration of the *Labyrinthodon salamandroides*, in the Secondary Island.   It is to be understood, however, that, with the exception of the head, the form of the animal is necessarily more or less conjectural.

### Nos. 19 & 20.—Labyrinthodon pachygnathus.

This name, signifying the Thick-jawed Labyrinthodon, was given

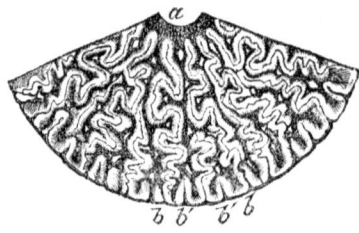

by its discoverer to a species of these singular Batrachia, found in the new red sandstone of Warwickshire, and which bears to the largest species the proportion exhibited by the head and fore-part of the body, as emerging from the water, for which parts alone the fossils hitherto discovered justify the restoration.*

Nos. 19 & 20. Section of Tooth of Labyrinthodon.
*a* Pulp-cavity : *b b* inflected folds of ossified capsule of tooth.

### Nos. 21 & 22.—Dicynodon.

In 1844 Mr. Andrew G. Bain, who had been employed in the construction of military roads in the colony of the Cape of Good Hope, discovered, in the tract of country extending northwards from the county of Albany, about 450 miles east of Cape Town, several nodules or lumps of a kind of sandstone, which, when broken, displayed, in most instances, evidences of fossil bones, and usually of a skull with two large projecting teeth.   Accordingly, these evidences of ancient animal life in South Africa were first notified to English geologists by Mr. Bain under the name of "Bidentals ; " and the specimens transmitted by him were sub-

* Conybeare, Geol. Trans., i. 388.

mitted at his request to Professor Owen for examination. The results of the comparisons thereupon instituted went to show that there had formerly existed in South Africa, and from geological evidence, probably, in a great salt-water lake or inland sea, since converted into dry land, a race of reptilian animals presenting in the construction of their skull characters of the crocodile, the tortoise, and the lizard, coupled with the presence of a pair of huge sharp-pointed tusks, growing downwards, one from each side of the upper jaw, like the tusks of the mammalian morse or walrus. No other kind of teeth were developed in these singular animals : the lower jaw was armed, as in the tortoise, by a trenchant sheath of horn. Some bones of the back, or vertebræ, by the hollowness of the co-adapted articular surfaces, indicate these reptiles to have been good swimmers, and probably to have habitually existed in water ; but the construction of the bony passages of the nostrils proves that they must have come to the surface to breathe air.

Some extinct plants allied to the Lepidodendron, with other fossils, render it probable that the sandstones containing the Dicynodont reptiles were of the same geological age as those that have revealed the remains of the Labyrinthodonts in Europe.

The generic name Dicynodon is from the Greek words signifying "two tusks or canine teeth." Three species of this genus have been demonstrated from the fossils transmitted by Mr. Bain.

The *Dicynodon lacerticeps*, or Lizard-headed Dicynodon, attained the bulk of a walrus ; the form of the head and tusks is correctly given in the restoration (No. 21) ; the trunk has been added conjecturally, to illustrate the strange combination of characters manifested in the head.

A second species, with a head so formed as to have given the animal somewhat of the physiognomy of an owl, has been partially restored at No. 22.

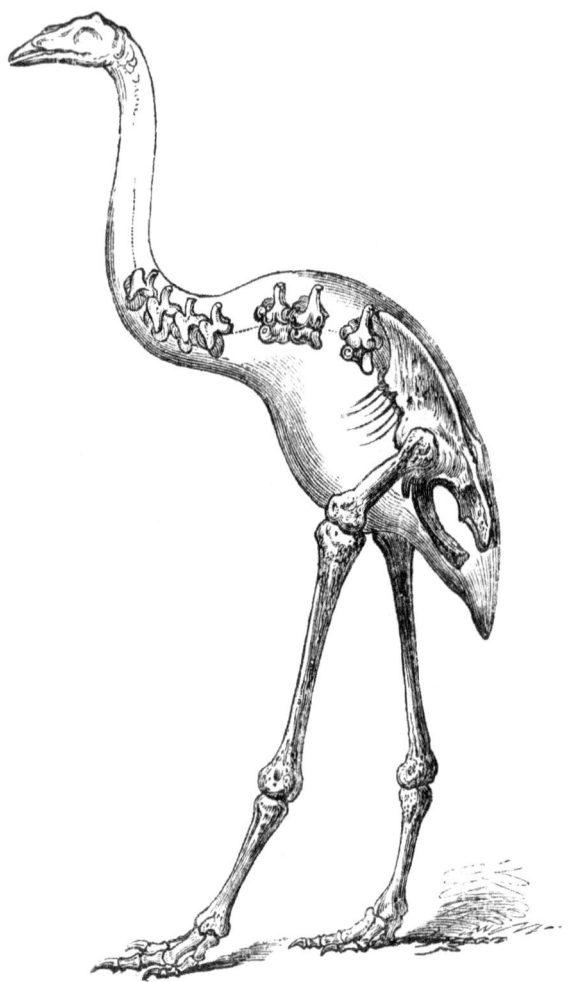

No. 8. Dinornis.

BRADBURY AND EVANS PRINTERS, WHITEFRIARS.

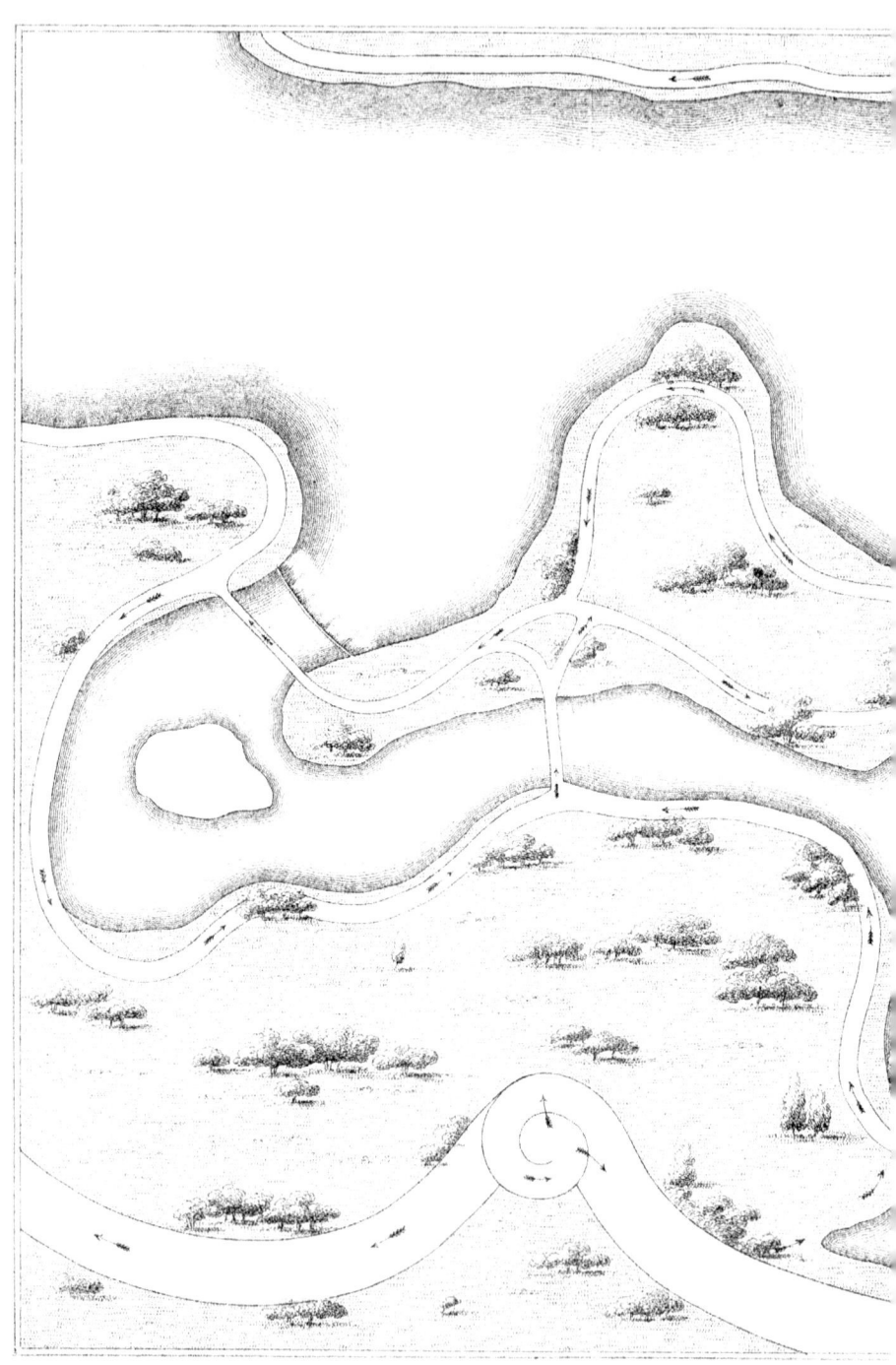

Published for the Crystal Palace Library By Bradbury & Evans, 11 Bouverie St.

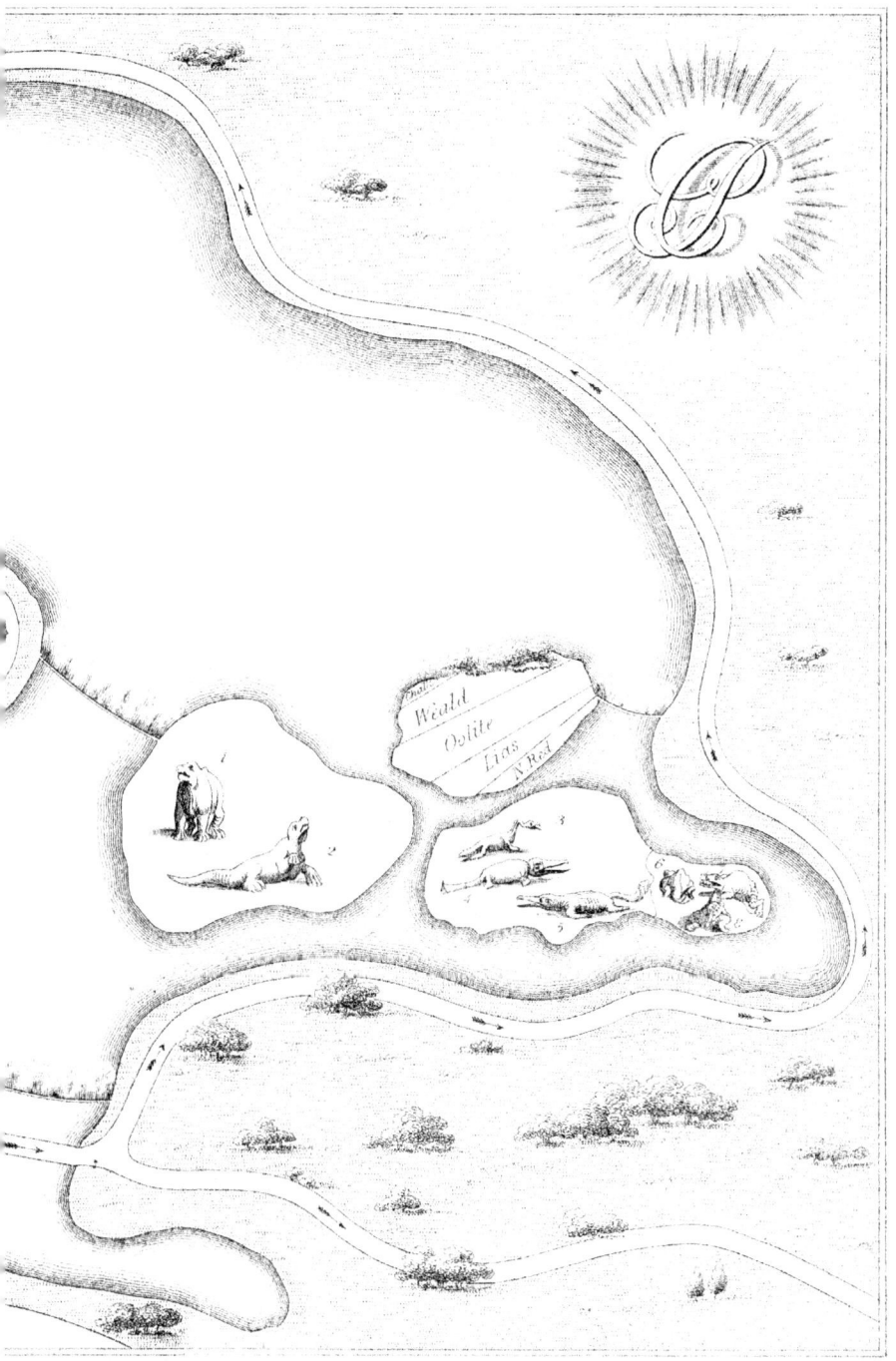

# Crystal Palace Guides series

The complete series will be available in facsimile editions. Check online.

# Euston Grove Press    www.EustonGrove.com